'Human Specialness':

The Historical Dimension & the Historicisation of Humanity

Peter Xavier Price

A catalogue record for this book is available from the British Library

ISBN: 978-1-907962-67-7

Published by Cranmore Publications

www.cranmorepublications.co.uk

The winning response to Neil Paul Cummins' 2010 book, *Is the Human Species Special?*, Peter Xavier Price's *'Human Specialness'* seeks to explore and challenge many of the epistemological suppositions undergirding the former's central ideas. Why, the author speculates, does the application of history play such a minor role in considerations of the supposed uniqueness of humanity? Likewise, can mankind's sense of its own historical nature pave the way towards a better informed and responsible future? Questions such as these, amongst many others, form the basis for this short book, in which humanity's eternal struggle to find inherent meaning in its surrounding world – as well as humanity's place *within* it – is reconsidered.

'Human Specialness': The Historical Dimension & the Historicisation of Humanity

Introduction

What is it about humanity that places it far above other life-forms? Why does it often perceive itself to be so unique when the natural world is teeming with biological anomalies? Perhaps even more tentatively, can humans truly claim to be the remedial agents destined to solve the current global environmental crisis? In Neil Paul Cummins' recent book, *Is the Human Species Special?*, the author sets out to address these very questions by speculating that mankind is indeed special because it represents the pinnacle of the evolutionary process.

Employing a radical thesis which bears a remarkable resemblance to the infamously distorted dictum of the Vietnam War (i.e., that of 'destroying the village in order to save it'),[1] Cummins suggests that mankind has reached a paradoxical stage in its development, whereby its imminent downfall may suddenly prove to be the means of its ultimate redemption. Thus, in this swashbuckling interpretation of the human response to environmental uncertainty, Cummins paints a picture of the human condition as seemingly analogous to the closing act in a grand, teleological narrative of biological endeavour and primordial purpose. 'Could it be', he speculates, 'that in order to fulfil its purpose and be the saviour of planetary life ... humanity had to believe that it was potentially the destroyer of planetary life?'.[2]

---

[1] The original expression, 'it became necessary to destroy the town to save it', was originally published by war journalist Peter Arnett on February 7 1968

[2] Neil Paul Cummins, *Is the Human Species Special?*, (Cranmore, 2010), p. 61

From the outset, it is important to note that Cummins' publication is an accomplished work – at once entertaining as it is erudite. The author clearly exhibits the full depth and range of his innate interdisciplinarity as he weaves seemingly disparate strands from his economic, environmental and philosophical background into a tightly argued and well-constructed piece. But what, we may be entitled to ask, are the inherent pitfalls to the bold thesis that he has constructed? Indeed, some may even believe that it falls short at the first hurdle. For how, they might argue, can the wiping out of a whole village constitute any sort of liberation for its inhabitants? Yet, as valid as this criticism may appear to be on the surface, it should be acknowledged that Cummins does in fact cover his tracks in this respect when he proposes that it is the *imminence* of the environmental disaster (rather than the purported disaster itself) that will ultimately ensure the planet's survival. Therefore, as far-fetched as the overarching argument may appear to be to some, it is simply wrong to accuse the author of outright contradiction.

This essay, then, is in large part an attempt to sketch out a far more convincing alternative to Cummins' arguments; but not, as may be expected, to what is essentially *the* central argument contained therein. In doing so, it aims to redeploy Cummins' ideas and to use them as a catalyst for further discussion; though, perhaps, in a direction that he mostly neglects or even ignores. At this initial stage, and in the interests of brevity, we may wish to describe this endeavour 'an assessment of the relative absence of *history* in Cummins' idiosyncratic account of human specialness'. For, appositely, this essay also seeks to highlight the importance of recognising humanity's unique *sense of its own historicity* – and, by extension, the decisive role that this must surely play in any adjudication of what it is to be an exceptional species. It is hoped, therefore, that we have already gone some way towards accounting for the choice phrases (i.e. 'historical dimension' and 'historicisation of humanity') which both comprise the frontispiece to this work. Nonetheless, what they mean in precise terms should become increasingly transparent as the essay develops. Suffice it to say that, having achieved this, we will then be in a much better position to review the suppositions undergirding Cummins' work.

I

Why, then, the historical dimension; any more so than, say, the purely scientific or the purely political? The answer to this question lies, firstly, in examining Cummins' choice of methodology; that is to say, the means by which he reaches his conclusion that humanity is in fact 'special'. Indeed, in the first half of the book, Cummins assembles an incremental argument essentially culminating in his belief that human beings are unique because they are a 'technological species'. This point is vitally important because technology, as far as the author is concerned, represents the sole means of minimising what he terms the 'deleterious effects of human-induced global warming'. However, it should be noted at this point that within this context, Cummins' use of the term 'technology' does not denote that which we already know to exist: such as mass recycling; the increased adoption of renewable energy sources; the carbon footprint campaign and so on. Rather, Cummins claims that these 'traditional' forms of environmental concern are ultimately incapable of achieving effective climatical reform, because humans have, in effect, transgressed to a degree far beyond their current restorative capabilities.

Thus, only by embracing what he considers to be the *'dominant force to environmental destruction'* – which is typified by mankind's acquisitive, materialistic nature, coupled with its relentless consumption and manipulation of resources – will the species then be forced to accept the imperativeness of acting upon the following quandary: either a) develop the future technology capable of regulating the earth's atmospheric temperature, thereby sustaining planetary life; or b) concede that this 'atmospheric temperature-controlling technology' will never be developed, to the utter ruin of all.[3] To be sure, in the preceding and concluding chapters, Cummins makes it abundantly clear which of the two he feels the human species is more likely to follow.[4]

As has already been intimated, criticism of Cummins' overriding theme resides not so much in the general principles of his theory, but, rather, in his relative neglect of what is, one suggests, an important constituent of how such a theory may be justifiably conceived of in the first place; namely, the aforementioned *historical dimension* to perceived notions of human specialness. But why exactly is this so? The fitting response to this

---

[3] Ibid., pp. 75-81; Cummins reaches these conclusions based upon the initial postulations of Sir James Lovelock's 'Gaia theory' at pp. 55-8

[4] Ibid., esp. pp. 82-94, 106-13

line of inquiry is, in the first instance, attributable to the fact that Cummins' ideas can be traced in a fairly linear trajectory from the bio-evolutionary thought of Charles Darwin (1809-82) down to the modern evolutionary consensus of the present day.[5] This is hugely significant, because though it is to be conceded that Cummins' theory is indeed unique, nonetheless, by virtue of the fact that it is essentially derivative in nature, it cannot help but be ensnared in what can only be described as an 'epistemological impasse'. What is exactly meant by this term may not appear to be immediately obvious, yet the argument that one is here attempting to make is that Cummins' thought (and he is certainly not alone in this respect) is actually predicated (like so many others) on two seem-

---

[5] Charles Darwin, *On the Origin of Species*, (London, 1859); *The Descent of Man*, 2 vols., (London, 1871); note Kim Sterelny's statement 'the development of evolutionary biology since 1858 is one of the great intellectual achievements of science' in his 'Philosophy of Evolutionary Thought' in *Evolution: The First Four Billion Years*, M. Ruse & J. Travis (eds.), (Cambridge, Harvard, 2009), p. 313

ingly innocuous – yet highly unsatisfactory – incongrui-ties. The first of these concerns the widespread, uncritical acceptance of scientific empiricism as the sole, pre-eminent means of reaching 'precise' conclusions. The second, pertaining particularly to the immediate case in hand, is the equally common assumption that within this trope, evolutionism is, as if beyond all reproach, the prerequisite means of gauging (or even validating) the human condition.[6]

This does not, of course, mean to suggest that Cummins is simply 'wrong' or 'inaccurate', simply because one may be either unconvinced by the Darwinian model (in particular), or modern forms of scientific inquiry (in general). On the contrary, it actually matters very little, at least for the purposes of this discussion, whether or not one chooses to believe in the supremacy of either 'science or religion', or even for that matter, of

---

[6] The term is most synonymous with Hannah Arendt's *The Human Condition*, (Chicago, Cambridge, 1958)

the separateness between 'fact and fiction' as night and day. Accordingly, this essay does not wish to digress (or regress) towards oversimplification, not least by pandering to the false dichotomies of the creationism versus evolutionism debate.[7] However, and crucially, what this essay *does* intend to do at this juncture is to propagate the following proposition: that is that, fundamentally-speaking, many of the widely pervasive epistemological assumptions permeating Cummins' overarching thesis have all transpired through a series of (seemingly unobserved) historical contingencies; sometimes great, and sometimes small. To some, this may not appear to be a particularly insightful observation. However, surely it must be insisted that historical contingencies deserve to be taken into much greater consideration, especially when the issue of 'human specialness' constitutes the current topic of debate.

---

[7] Cummins acknowledges creationism in *Human Species*, pp. 49-50

Of course, Cummins is not so naïve as to suggest that it is only now, in the wake of greater advances in the sciences, that the question of human specialness has only very suddenly risen to central importance. In addition, this also means that he simply cannot be – and definitely is not – oblivious to the innate historicity clearly evident in humanity's sense of self. Indeed, even as early as in the introduction, the author states that human specialness is an issue 'which humans have been pondering for many millennia'.[8] Yet, despite all this, it is still reasonable to maintain that criticism of Cummins' approach remains valid because it does not appear to take seriously enough the *extent* to which mankind seems perpetually prepared to grapple with its own past.

As a result, many important questions often seem to be overlooked in his account of human specialness, such as the question '*why*', for example, '*have humans pondered the issue of human uniqueness for so long?*'; or

---

[8] Ibid., p. 9

even '*why is scientific empiricism so often considered to be the pre-eminent means of reaching precise conclusions?*'. Yet another question pertinent to the point under examination might also be: '*why does contemporary wisdom claim that contingent factors are so important in explanations of historical causation and/or evolution?*'. Admittedly, Cummins does eventually tread very lightly upon this latter theme. But again, the issue is afforded somewhat scant attention, merely, it seems, because the author does not appear to think that it offers much by way of a challenge to his central thesis.[9] This may well be true enough, if discussed only very briefly and, specifically, within the bio-evolutionary idiom. But there are certainly many other ways in which the issue can be better explored.

---

[9] Ibid., pp. 100-1

II

Talk concerning 'historical contingencies' or 'epistemological impasses' is not in any way intended to be a deliberate exercise in esotericism. Rather, such language has been utilised merely as a practicable (though perhaps slightly surreptitious) means of emphasising just how easily we, as humans, take for granted the numerous conditions under which the attainment of knowledge is possible. For the purposes of this debate, it is important, then, to underline once again that which we are most concerned with in this essay: namely, the *historical dimension* to notions of human specialness. For, in contradistinction to Cummins' bio-evolutionary model, it is now proposed that it is in the application of *history* that we may, perhaps, derive one of the most significant forms of possible knowledge (or, of *possible forms of knowledge*). In order to illustrate the point, let us turn momentarily to Cummins' later interesting admittal of the intellectual debt he feels he owes to the German Romanticist, Friedrich Hölderlin (1770-1843). For, by doing so, it becomes increasingly apparent that the author does not merely recount the philosophical influence of the early-nineteenth-century thinker, but, rather, that he actually attempts to use Hölderlin's thought as a vehicle for 'canonical' explanation: firstly, by expounding the veracity of Darwin's ideas in relation to Hölderlin's; and secondly, by then doing the same in relation to his own:

> ... Darwin's theory of evolution, which emerged shortly after Hölderlin's time, gives a view of evolutionary processes that is incompatible with Hölderlin's view ... [nonetheless] it is clear that this Darwinian based objection does not invalidate the view of Hölderlin, or the reinterpretation of them presented in this paper.[10]

---

[10] Ibid., Appendix A: 'Human nature, cosmic evolution and modernity in Hölderlin' at p. 143

What this excerpt serves to illustrate is that Cummins' use of history, his engagement with history—indeed, *his very historicity*—is exemplary of the fact that we are all, in effect, in constant conversation (or dialogue) with the past. As a result, whether we choose it or not (if, indeed, we are even aware of it in the first place), we are all unwittingly, inescapably and paradoxically, both its pupil *and* teacher; for mankind's capacity to record its own history is *itself* a product of the historical imagination.[11] Indeed, if we further expand upon the idea of history as a uniquely human activity, then we will find that it is fundamentally diverse in its range and scope and that it has, throughout the ages, always been utilised in a

---

[11] R. G. Collingwood, *The Idea of History*, (Oxford, 1946); for further discussion on eschewals of historical 'canonicity', see *Philosophy, Politics, and Society*, 2nd series, P. Laslett and W. G. Runciman (eds.), (Oxford, 1962); John Dunn, 'The Identity of the History of Ideas', *Philosophy* 43 (1968), pp. 85-104; Quentin Skinner, 'Meaning and Understanding in the History of Ideas', *History and Theory* 8, (1969), pp. 3-53; 'General Introduction' to both *Economy Polity and Society* & *History Religion and Culture*, S. Collini, R. Whatmore and B. Young (eds.), (Cambridge, 2000)

large variety of ways. From historical narrative, analysis and myth-making to drama, poetry and biography; history can be (and often has been) reshaped into epic-storytelling, investigative journalism or even political inquest; and ever has it been susceptible to the ruses of polemicists, propagandists and writers of jeremiads.[12]

Why, then, it should be asked, is Cummins' appreciation of 'human historicity' either, at best, simply de-emphasised, or at worst, completely overlooked? Indeed, let us recall for a moment the underlying crux of his conception of human specialness – i.e., that it is a 'technological species'. For, at numerous points through-out the book, the author continually asserts that the development of the human species – from 'hunter-gathering' to 'technological' – is itself a part of the evolutionary process.[13] Yet, by doing so, Cummins'

---

[12] John Burrow, *A History of Histories*, (Penguin, 2007), esp. 'Introduction', at pp. xiii-xix

[13] Cummins, *Human Species*, pp. 18, 62, 67-9, 94

awareness of history appears at all times, again, to be relegated to that of a bit-part role. In particular, the two 'historical stages' which he believes to be 'integral' to the development of our 'technological species' – both the 'scientific revolution' and the 'industrial revolution' – end up acting as mere forerunners to the main event: for both, he generalises, 'had to happen *somewhere* in particular ... As it turned out, it was in Western Europe that the human species *became* technological'.[14]

Such flagrant disregard for the rich complexities inherent within such weighty terms as 'scientific revolution' or 'industrial revolution' can only lead one to the conclusion that the author simply does not consider history to be a particularly valuable tool of explanation; certainly not, at least, in comparison to the extent that he believes both evolutionism and empiricism are. This approach, one contends, is wholly unsatisfactory, because it infers that history is secondary to evolutionism or

---

[14] Ibid., p. 72, emphases added

empiricism, when it should be considered, intellectually-speaking, at least on a par. Indeed, if Cummins fails to acknowledge the legitimacy of this claim, then it would surely appear to be the case that he also completely disregards the valuable notion that both evolutionism and empiricism are, in fact, *historical constructs* themselves. Similarly, even if he does not entirely disagree with this point of view, the fact that throughout the book he fails to fully engage with its complex and manifold implications means that many doubts may be justifiably raised about what may be termed the 'structural integrity' of the thesis he purports.

It is for these very reasons, then, that Cummins' apparent diffidence to what we have chosen to call the *historicisation of humanity* greatly weakens so much of what it is he is attempting to put across in his arguments. This having been said, a belated yet vitally important subsidiary concern of this essay is to establish precisely how (and why) this may be accounted for; in particular, by closely examining some of the predominant issues which certainly *were* of genuine concern to many of those who were alive during the so-called 'scientific' and 'industrial revolutions'. Furthermore, by momentarily overlooking the fact that hegemonic interpretations of Darwinism simply do not stand up to close, historical scrutiny,[15] we will instead turn our attentions to an earlier time period, when the concept of evolutionism did not even yet exist. For, by doing so, it will also be seen that general, teleological interpretations of humanity's supposed purposiveness were, in fact, highly conspicuous by their absence. Indeed, history had first to be contended with.

---

[15] See Burrow, *Evolution and Society*, (Cambridge, 1966); *The Crisis of Reason: European Thought, 1848-1914*, (Yale, 2000), esp. Chap. 2, 'Social Evolution and the Sciences of Culture', pp. 68-108

III

The case can initially be made that significant portions of Cummins' thought correspond more closely with historically *cynical*, rather than *optimistic*, interpretations of human nature; especially when we pause to consider the author's statement that human existence is, in effect, analogous to planetary life's inexorable *striving for survival*.[16] Indeed, the Dutch-Jewish Pantheist, Baruch Spinoza (1632-77), stated much the same himself nearly three and a half centuries ago when he declared that mankind simply cannot, and will not, deprive itself of those things that it judges to be the most conducive to its welfare. It is a principle 'inscribed so firmly in the human breast', he declared, that it constitutes one of those 'immutable' human truths that 'nobody [can]

---

[16] At *Human Species*, p. 91

24

ignore'.[17] The British thinker, John Locke (1632-1704), argued a similar case two decades later, when he stated that mankind's perpetual resolve to ensure its own survival – its constant 'uneasiness' as he called it – constituted the 'chief, if not only spur to human industry and action'.[18] Both initial examples are extremely instructive because they illustrate the fact that many thinkers of the seventeenth-century began to speculate that the basis of human nature could simply be adumbrated by its instinct for 'self-preservation'. As a direct corollary, this meant that the perceived arbitrariness of 'self-preservation' might, in fact, even be able to account

---

[17] Baruch Spinoza, *Tractatus Theologico-Politicus* [1670], Emilia Giancotti Boscherini (ed.), (Torino, 1972), I, p. 472

[18] John Locke, *An Essay concerning Human Understanding*, (London, 1690), Book II, Chapter XX

for much of the conflict which evidently comprised a great deal of human history.[19]

Infamously, when Bernard de Mandeville (1670-1733), originally a Dutch-physician with an acute eye for satirical controversy, settled in England in the 1690s, he brought with him an abrasive reformulation of such Spinozist and Lockean ideas, whereupon a wholly *disparaging* account of humanity's supposed specialness came suddenly, and contentiously, to the fore. Comparable to Thomas Hobbes (1588-1679) in his dismissal of the notion that man was the *zōon politikon*,[20] Mandeville instead cunningly sought to dispel the myth that either reason or religion was capable of controlling the irresisti-

---

[19] The latest, and most immediately important to the thinkers of the era, were the pan-European 'Wars of Religion' of the sixteenth-and seventeenth-centuries; the widespread influence of Thomas Hobbes is particularly instructive, esp. his *Leviathan* [1651], Richard Tuck (ed.), (Cambridge, 1966)

[20] Meaning 'the political animal', which is, of course, the most famous dictum of Aristotle's *Politics*

ble ebb and flow of humanity's baser passions.[21] For him, then, the notion of self-preservation was really a byword for the ineradicable compulsions of 'self-love'. Moreover, he believed that self-love, when stripped of its idealistic undertones, was in fact merely tantamount to 'avariciousness'.

Yet, contrary to what one might think, Mandeville did not necessarily consider this state of affairs to be problematic. For even though the Dutch-native observed that modern commercial societies seemed increasingly to be mired in vice, licentiousness, pride and deceit,

---

[21] Scholarship on Mandeville is exhaustive. The best study still appears to be, E. J. Hundert, *The Enlightenment's 'Fable': Bernard Mandeville and the Discovery of Society*, (Cambridge, 1994); see also, Hector Monroe, *The Ambivalence of Bernard Mandeville*, (Oxford, 1975); Thomas A. Horne, *The Social and Political Thought of Bernard Mandeville*, (London, Macmillan, 1978); M. M. Goldsmith, *Private Vices, Public Benefits*, (Cambridge, 1985), Dario Castiglione, 'Mandeville moralised', *Annali Della Fondazione Luigi Einaudi* 17 (1983), pp. 239-290; 'Considering Things Minutely: Reflections on Mandeville and the Eighteenth-Century Science of Man', *History of Political Thought* 7 (1986), pp. 463-488.

nonetheless, he also argued that this could be more than justified by the intrinsic *human flourishing* (and overall peace-keeping) that it thereby established. Thus, in the most famous of his works *The Fable of the Bees: or, Private Vices, Public Benefits* (1714-23), Mandeville alleged that it was simply far more realistic and expedient for political authorities to concede that modern commercial society flourished best through the combined avarice, pride and utility of its inhabitants. Only in so doing, he argued, could the State anticipate being 'flattered in peace, and feared in war' – for 'Moral virtues', he claimed, were nothing more than the 'Political Offspring which Flattery begot upon Pride':

> Thus every part was full of vice/ yet the whole mass was a paradise ... whilst luxury/ employed a million of the poor, and odious pride a million more/ envy itself and vanity, were ministers of industry ... Thus vice nursed ingenuity/ which joined with time and

industry ... to such a height, the very poor/ lived better than the rich before.[22]

Unsurprisingly, given the fact that the early eighteenth-century was still an overwhelmingly theocratic age, Mandeville's acerbic pessimism was largely condemned, and he was chastised by many as an outright hedonist and debauchee. However, and crucially, the recalcitrant legitimacy of many of his claims simply could not be ignored by either his contemporaries or the subsequent generation, and it is in this way that he inadvertently contributed to that broad intellectual movement later known as the 'Scottish Enlightenment'. For although many earlier responses to the cynicism of Mandeville had attempted to redefine the relationship between private interest and public good as one of moral

---

[22] Bernard de Mandeville, *The Fable of the Bees: or, Private Vices, Publick Benefits* [1723], F.B. Kaye ed., (Oxford, 1924), pp. 24-6, I: 51

compatibility rather than friction,[23] it was the illustrious David Hume (1711-76) who first recognised and celebrated the fact that Mandeville had effectively 'put the science of man on a new footing'.[24]

To be sure, Hume's development of the Dutch-physician's ideas prompted the Scotch-philosopher and historian to conclude that mankind was, indeed, essentially ruled by the Passions.[25] By this means, he further concluded that neither religion nor reason – including many aspects of 'Newtonian' empiricism (believed by

---

[23] Predictably, these were mainly steeped in Christian notions of morality, see esp. William Law, *Remarks upon a late Book* [1724] & *A Serious Call* [1728], *Works*, vols. II, IV, (London, 1762); Joseph Butler, *Fifteen Sermons*, (London, 1726), esp. Sermons XI & XII

[24] David Hume, *A Treatise of Human Nature* [1739], L. A. Selby-Bigge (ed.), (Oxford, 1896), I. 'Introduction'; see also Adam Smith's statement that Mandeville may have been a 'coarse and rustic' writer, yet at the same time 'lively and humorous' in his *The Theory of Moral Sentiments* [1758][hereafter, *TMS*], Knud Haakonssen (ed.), (Cambridge, 2002), pp. 53, 363-4

[25] Hume, *Treatise*, Vol. II, Book II. 'Of the Passions'

many to be a chief characteristic of both the 'scientific revolution' and/or the English enlightenment) – was truly capable of accounting for the establishment of modern civil society.[26] In turn, and in ways which are probably beyond the scope of this essay to fully explore, this facilitated the era's growing concern (or penchant) for 'conjectural history'—that branch of historical inquiry most concerned with attempting to account for mankind's putative ascent from 'rudeness, rusticity and barbarism' to 'commerce, refinement and civilisation'.[27]

---

[26] Hume, *Essays, Moral, Political and Literary*, [1742-**1777]**, Eugene F. Miller (ed.), (Indianapolis, 1987); let us not also forget the influence that he had on the thought of Immanuel Kant, esp. *The Critique of Pure Reason* [1781], (Cambridge, 1998)

[27] Well-known examples include Adam Ferguson, *An Essay on the History of Civil Society* [1767], Duncan Forbes (ed.), (Edinburgh, 1966); William Robertson, 'A View of the Progress of Society in Europe …' in *The History of the Reign of the Emperor Charles V*, (London, 1769); Henry-Home Lord Kames: *Sketches of the History of Man* [1774], 4th ed., 4 vols., (Edinburgh, 1788). It has not escaped notice that these examples appear to be overwhelmingly, and perhaps unjustifiably, British (or Scottish)-centric in nature. Yet

What is more, this (predominantly Scottish) programme of historical, moral and civic analysis even acquired a title of its own – 'the four-stages theory' – in light of the fact that it attempted to delineate the possible development of human society through a series of well-defined, successive stages: from hunting, pastoral and agricultural to commercial.

The establishment of conjectural history, alongside the 'four-stages theory', effectively paved the way for a readier reception of the idea that historical contingency

---

continental examples of the contemporary trend for conjectural history can also be seen in, for example, large swathes of the *Encyclopédie, ou dictionnaire raisonné des sciences, des arts et des metiers* [*Encyclopaedia or a Systematic Dictionary of the Sciences, Arts and Crafts*], Denis Diderot & Jean le Rond d'Alembert (eds.), (Paris, 1751-72); Jean-Jacques Rousseau, *Discours sur l'origine et les fondements de l'inégalité parmi les homes* [*Discourse on the Origin and Basis of Inequality among Men*], [1754-5]; Kant, *Idea for a Universal History from a Cosmopolitan Point of View* [1784], trans. A. W. Wood in 'The Cambridge Edition of the Works of Immanuel Kant', *Anthropology, History, and Education*: trans. & eds. G. Zöller & R. B. Louden, (Cambridge, 2007)

– in particular, the development of civil society running in tandem with a series of unintended consequences – was, in fact, the most accurate means of accounting for contemporary circumstances. Moreover, because this also heavily implied that human nature could not simply be reduced to all-enveloping universal principles, there further developed the systematic propagation of what was to become known as the 'science of legislation' – i.e., the means by which wise and responsible governments might be able to guard against unforeseen historical circumstances, firstly by establishing, and then maintaining, the conditions necessary for the promotion of peace, stability and security for all.[28]

---

[28] See, III. 'Natural jurisprudence and the science of legislation' in *The Cambridge History of Eighteenth-Century Political Thought*, Mark Goldie & Robert Wokler (eds.), (Cambridge, 2006). For later developments, especially in the nineteenth-century, see Stefan Collini, Donald Winch & John Burrow, *That Noble Science of Politics*, (Cambridge, 1983)

Indeed, no more keenly was this enterprise felt than in the sphere of political economy; a discipline which can be best described as the attempt to ally the theories of the 'science of legislation' with the ever-increasingly, economically 'self-aware' concerns of the modern state.[29] Although modern economics has since garnered a reputation for being a somewhat cold, detached and mechanistic discipline, it is worth considering that *political economy*, by contrast, was at its root intensely aware of its own moral (and in some cases amoral) underpinnings. Adam Smith (1723-90), for example – the stylised father of the liberal economic tradition – phlegmatically foresaw that the monotonous drudgery of

---

[29] See esp. Josiah Tucker, *The Elements of Commerce and the Theory of Taxes*, (Privately Published, 1755); *An Essay on Trade* [1749], *A Brief Essay on the Advantages and Disadvantages which respectively attend France and Great Britain, with regard to Trade*, (3rd Ed., London, 1753); Smith, *Wealth of Nations*, (London, 1776); *Wealth and Virtue: The Shaping of Political Economy in the Scottish Enlightenment*, I. Hont and M Ignatieff (eds.), (Cambridge, 1983); Winch, *Riches and Poverty: An Intellectual History of Political Economy in England 1750-1834*, (Cambridge, 1996)

the division of labour, congruent with the soullessness of all 'mercenary exchange[s]',[30] might in fact 'dehumanise' the working population in ways that Karl Marx (1818-83) was to fully, and famously, exploit in his own historical-materialist adaptation of conjectural history. Nonetheless, for others, most notably a slightly older contemporary of Smith, the Anglican Dean of Gloucester Josiah Tucker (1712-99), the growing 'science' of political economy instead constituted the means by which humanity's material, and thereby spiritual, condition might be exponentially improved. Indeed, it was his staunch belief that it was, in fact, God's providential plan for mankind to cultivate sophisticated and *meaningful* forms of social relations, so that they might procure the common enjoyment of all the fruits of the earth, no matter where produced:

---

[30] Smith, *TMS*, Part II, II, Chap. III. 2, p. 100

What *general* Rule can we pursue for the *mutual* Benefit of Mankind? And how are the Ends both of Religion and Government to be answered, but by the System of universal Commerce?—Commerce, I mean, in the large and extensive Signification of that Word; Commerce, as it implies a general System for the useful Employment of our Time; as it exercises the particular Genius and Abilities of Mankind in some Way or other, either of Body or of Mind, in mental or corporeal Labour, and so as to make Self-interest and Social coincide. And in pursuing this Plan, it answers all the great Ends both of Religion and Government; it creates social Relations; and it serves as a Cement to connect together the Religious and Civil Interests of Mankind.[31]

---

[31] Tucker, *Seventeen Sermons on Some of the Most Important Points on Natural and Revealed Religion, Respecting the Happiness Both of the Present and of a Future Life*, (Gloucester, 1776), pp. 131-9

Of course, it is all too easy for the modern (or perhaps even post-modern), sceptical reader to assert that Tucker's (and to some extent, even Smith's) vision is utterly incompatible with the uncompromisingly secular and technologically-advanced preoccupations of the Western world today. Yet it is worth remembering that political economy, whilst a somewhat outdated discipline by today's standards, is still considered by some to be the first and most definitive social science of our era, and that its underlying principles therefore remain the foremost paradigm within which the modern world operates. Even more significantly, as we have just touched upon, the sombre moral discourses which initially contributed to its development throughout the eighteenth-century deserve to be taken into much greater consideration at present; even more so when we consider the ticking time bomb that is the issue of global warming and/or climate change. For how else do we propose to account for times of relative scarcity for many; of food shortages and rising unemployment; or, indeed, for the fact that geopolitical tensions are continually exacerbated by mounting global economic – and, of course, environmental – uncertainty?[32] Indeed, the fact that these seemingly disparate issues are, in fact, patently interconnected and wholly dependent upon one another (whether negatively or, potentially, positively), means that the *moral* perspective should always remain prominent, which, unfortunately, appears rarely to be the case.

What this exercise in historical exposition has attempted to show, then, is that grandiose 'master narratives' of the supposed uniqueness of humanity (in which axioms such as secularisation, liberalism and evolutionism tend, often, to be labelled as the intellectual *telos* of mankind)[33] simply do not offer much by way of

---

[32] See *The Economic Limits to Modern Politics*, John Dunn (ed.), (Cambridge, 1990); Istvan Hont, *Jealousy of Trade: International Competition and the Nation State in Historical Perspective*, (Cambridge, MA, 2005)

[33] In other words, 'Whiggish history', see Herbert Butterfield's *The Whig Interpretation of History*, (London, 1931)

practicable solutions to contemporary problems; especially when, as is so often the case, there remain a plethora of often complex historical contingencies which are yet to be satisfactorily unravelled and exposed. This idea is certainly nothing new. Indeed, Edmund Burke (1729-97), on the eve of the Terror in Revolutionary France, keenly observed that abstract, sanguine notions of universality, moral certainty and natural rights could just as easily degenerate into naïve speculation, mischievous abstraction or even outright skulduggery and bloodshed; and the subsequent events of 1793-4 certainly seem to vindicate much of what it was he had to say in his *Reflections on the Revolution in France* (1790).[34] If we are to utilise the analogy further for a moment (though, of course, in a vastly different context), it will be seen that criticism of Cummins' approach throughout this essay now appears to be reaching its climax. For, in ironically

---

[34] Edmund Burke, *Reflections On The Revolution In France* [1790], Conor Cruise O'Brian (ed.), (Penguin, 1973)

propagating a thesis which remains largely dependent upon the traditions of empiricism, it is with some irony that his is, in fact, a decidedly *speculative* and *metaphysical* response to potential environmental catastrophe – and one that Burke would certainly recognise as such if he were alive today.

Indeed, Cummins' shortcomings are even further compounded by his exploitation of a number of schemes within his thesis which, as we have shown, are demonstrably *historical*, and yet do not appear to be *historically* accounted for. For it surely cannot have escaped notice that Mandeville's early account of wealth-creation, via the paradox of 'unsocial sociability', bears more than a passing resemblance to the author's bio-evolutionary (or even quasi-eschatological) account of the potentially redemptive qualities of 'fallen' man. A similar case may even be inferred by his adoption of decidedly Malthusian concepts, about which, again, there appears to be no acknowledgment at all.[35] Yet, even more significantly, Cummins' account of what he calls the 'trajectory of human evolution from hunter-gatherer to technological society'[36]—indeed, the very thread upon which his whole argument is based—appears, in truth, to be little more than the eighteenth-century Scottish 'four-stages-theory', albeit in slightly modified form. Had Cummins acknowledged this interesting fact, he might even have reached the conclusion that we may now be entering (or already find ourselves in) a quinquennial, climatical phase of a potential 'five-stage theory', replete with its own conundrums and challenges. Since he does not, it is with deep regret that the author seems so unable to construct a thesis containing greater reference to, and perhaps greater reverence for, crucial historical antecedents. For,

---

[35] Cummins' adoption of 'Malthusian' themes concerning population are at p. 73; *cf.* Thomas R. Malthus, *An Essay on the Principles of Population* [1798-1826], Patricia James (ed.), 2 vols., (Cambridge, 1989)

[36] *Human Species*, p. 94

if he had done so, it certainly would have been that much more difficult to dispute so many of the arguments contained therein.

Conclusion

The predominant concern of this essay has been the explication of humanity's sense of its innate historicity, alongside the extent to which this can be said to support notions of human specialness. In firstly constructing an argument which seeks to dispel the myth of epistemological certainty, a *theoretical alternative* has been devised which, if only temporarily, endeavours to place *human history* at the forefront of speculative inquiry. By this means, it has since been established that the human species represents the single form of planetary-life capable of engaging in this sort of activity. Consequently, the further claim that the *historicisation of humanity* might, in fact, contribute towards conceptions of human uniqueness appears to be truly genuine. Moreover, by applying these maxims to some of the most important socio-philosophical debates of a particular historical period, it has even been possible to historically-observe how many thinkers of *that* era used their burgeoning sense of *their own historicity* in order to account for, and, in places, better their current circumstances. For this reason, would it not be prudent for politicians, policy-makers and the general public alike, to do the same today?[37]

Detailed contextual analysis of the type that one has sought to provide throughout this essay offers a telling indication of the extent to which history can, and does, contribute to the attainment of human knowledge. To be sure, even the bio-evolutionary model is a mere

---

[37] Gordon J. Schochet, 'Why should history matter? Political theory and the history of discourse', in *The Varieties of British Political Thought 1500-1800*, J. G. A. Pocock, G. J. Schochet & L. G. Schwoerer (eds.), (Cambridge, 1993)

extension of this fact, even if its proponents may prefer to claim that it is a purely theoretical (that is to say, scientific and/or empirically based) enterprise. This, it must be stressed again, is not to insist that history constitutes *the* superlative means of *all* forms of intellectual inquiry; for this is indeed an outright absurdity. However, the fact that it is an academic discipline (or even leisure interest) which remains steadfastly ubiquitous, and yet highly transpositive in nature, means that it is constantly able to reinvent itself – just as the human species appears particularly adept at doing. Indeed, this is a quality to be celebrated and nurtured. It is to be hoped, therefore, that the very malleability of the human species may once again come to the fore as it attempts to deal positively with the growing threat posed by global, environmental tragedy.

Bibliography

Primary

Burke, E: *Reflections On The Revolution In France And On The Proceedings In Certain Societies in London Relative To That Event* [1790]: Conor Cruise O'Brien (ed.), (Penguin Books, Great Britain, 1973).

Butler, J: *Fifteen Sermons Preached at the Rolls Chapel*, (London, 1726).

Cummins, N. P.: *Is the Human Species Special? Why human-induced global warming could be in the interests of life*, (Cranmore, Reading, England, 2010).

Encyclopédie, ou dictionnaire raisonné des sciences, des arts et des metiers [*Encyclopaedia or a Systematic Dictionary of the Sciences, Arts and Crafts*], Diderot, D & d'Alembert (eds.), (Paris, 1751-72).

Darwin, C: *On the Origin of Species by Means of Natural Selection, or the Preservation of Favoured Species in the Struggle for Life*, (London, 1859).

———. *The Descent of Man, and Selection in Relation to Sex,* 2 Vols., (London, 1871).

Ferguson, A: *An Essay on the History of Civil Society* [1767], Duncan Forbes (ed.), (Edinburgh University Press, 1966).

Hobbes, T: *Leviathan, or The Matter, Forme, & Power of a Common-wealth   Ecclesiasticall and Civill* [1651], Richard Tuck (ed.), (Cambridge University Press, 1996).

Hume, D: *An Enquiry Concerning the Principles of Morals* [1751], (ed. New York, Prometheus, 2004).

———. *A Treatise of Human Nature* [1739], L. A. Selby-Bigge (ed.), 3 Vols., (Oxford: Clarendon Press, 1896).

———. **Essays Moral, Political and, Literary** *[1741-1777],* edited and with a Foreword, Notes, and Glossary by Eugene F. Miller, with an appendix of variant readings from the 1889 edition by T.H. Green and T.H. Grose, (rev. ed.), (Indianapolis, Liberty Fund, 1987).

Kames, H-H: *Sketches of the History of Man* [1774], 4th ed., 4 Vols., (Edinburgh, 1788).

Kant, I: *Idea for a Universal History from a Cosmopolitan Point of View* [1784],   Translated by A. W. Wood in 'The Cambridge Edition of the Works of Immanuel Kant', *Anthropology, History, and Education*, Translated and Edited by G. Zöller & R. B. Louden, (Cambridge, 2007).

———. *The Critique of Pure Reason* [1781], The Cambridge edition of the Works of Immanuel Kant, Translated and Edited by Paul Guyer & Allen W. Wood, (Cambridge University Press, 1998).

Law, W: *The Works of the Reverend William Law, A.M.*, 8 Vols., (London, 1762).

Locke, J: *An Essay concerning Human Understanding*, (London, 1690).

Malthus, T. R: *An Essay on the Principles of Population* [1798-1826], Patricia James (ed.), 2 Vols., (Cambridge University Press, 1989).

Mandeville, B: *The Fable of the Bees: or, Private Vices, Publick Benefits*, [1723], F.B. Kaye ed., (Oxford University Press, 1924).

Robertson, W: 'A View of the Progress of Society in Europe from the Subversion of the Roman Empire to the Beginning of the Sixteenth Century' in *The History of the Reign of the Emperor Charles V*, (London, 1769).

Rousseau, J-J: *Discours sur l'origine et les fondements de l'inégalité parmi les homes* [*Discourse on the Origin and Basis of Inequality among Men*], [1754-5].

Smith, A: *An Inquiry into the Nature and Causes of the Wealth of Nations*, (London, 1776).

———. *The Theory of Moral Sentiments* [1758], Knud Haakonssen (ed.), (Cambridge, 2002).

Spinoza, B: *Tractatus Theologico-Politicus* [1670], ed. Emilia Giancotti Boscherini, (Torino, Einaudi, 1972).

Tucker, J: *A Brief Essay on the Advantages and Disadvantages which Respectively Attend France and Great Britain with Regard to Trade* [1749], (3rd Ed., London, 1753).

———. *Seventeen Sermons on Some of the Most Important Points on Natural and Revealed Religion, Respecting the Happiness Both of the Present and of a Future Life*, (Gloucester, 1776).

———. *The Elements of Commerce and the Theory of Taxes* (Privately Published, 1755).

Secondary

Advances in Intellectual History, Richard Whatmore & Brian Young (eds.), (Palgrave, Macmillan, 2006).

Arendt, H: *The Human Condition*, (Chicago University Press: Cambridge University Press, 1958).

Berlin, I: *The Crooked Timber of Humanity*, Henry Hardy Ed., (Princeton University Press, 1998).

Burrow, J: *A History of Histories: Epics, Chronicles, Romances and Inquiries from Herodotus and Thucydides to the Twentieth Century*, (Penguin, Great Britain, 2007).

———. *A Liberal Descent: Victorian Historians and the English Past*, (Cambridge University Press, 1981).

———. *Evolution & Society: A Study in Victorian Social Theory*, (Cambridge University Press, 1966).

———. *The Crisis of Reason: European Thought, 1848-1914*, (Yale University Press, 2000).

Butterfield, H: *The Whig Interpretation of History*, (London, 1931).

Castiglione, D: 'Considering Things Minutely: Reflections on Mandeville and the Eighteenth-Century Science of Man', *History of Political Thought* 7 (1986), pp. 463-488.

———. 'Mandeville moralised', *Annali Della Fondazione Luigi Einaudi* 17 (1983), pp. 239-290.

Collingwood, R.G.: *The Idea Of History*, (Oxford University Press, 1946).

Collini, S., Winch, D. & Burrow, J: *That Noble Science of Politics*, (Cambridge University Press, 1983).

Cragg, G.R: *Reason and Authority in the Eighteenth Century*, (Cambridge University Press, 1964).

David Hume's Political Economy, Schabas and Wennerlind, (eds.), (Routledge, London, 2008).

Dunn, J: 'The Identity of the History of Ideas' *Philosophy*, Vol. 43, No. 164 (Apr., 1968), pp. 85-104.

Economics and Interdisciplinary Exchange, Guido Erreygers (ed.), (London, Routledge, 2001).

Economy, Polity, and Society & *Religion, Culture & History: British Intellectual History 1750-1950*, S. Collini, R. Whatmore & B. Young Eds., (Cambridge University Press, 2000).

Evolution: The First Four Billion Years, M. Ruse & J. Travis (eds.), (Cambridge, Harvard, 2009).

Forbes, D: *Hume's Philosophical Politics*, (Cambridge University Press, 1975).

Gauthier, D.P..: *The Logic of Leviathan: The Moral and Political Theory of Thomas Hobbes*, (Oxford University Press, 1969).

Goldsmith, M. M.: *Private Vices Public Benefit: Bernard Mandeville's Social and Political Thought*: (Cambridge University Press, 1985).

Haakonssen, K., Whatmore. R., De Lucca, J-P., 'Essay Reviews: Commerce and Enlightenment', *Intellectual History Review*, 18 (2) 2008, pp. 283-306.

Hont, I: *Jealousy of Trade: International Competition and the Nation State in Historical Perspective*, (Cambridge, MA, 2005).

Horne, T.A: *The Social and Political Thought of Bernard Mandeville*, (London, Macmillan, 1978).

Hundert, E.J.: *The Enlightenment's 'Fable': Bernard Mandeville and the Discovery of Society*: (Cambridge University Press, 1994).

Koehn, N. F: *The Power of Commerce: Economy and Governance in the First British Empire*, (Ithaca and London, Cornell University Press, 1994).

Malthus, medicine and morality: 'Malthusianism' after 1798, Brian Dolan, (ed.), (Atlanta, GA, 2000).

Monroe, H: *The Ambivalence of Bernard Mandeville*, (Oxford University Press, 1975).

Philosophy, Politics, and Society, 2nd series, P. Laslett and W. G. Runciman (ed.), (Oxford, 1962).

Pocock, J. G. A: *Barbarism and Religion*, 4 Vols. (Cambridge, 1999-2005); Vols. 5 & 6 forthcoming (2010 -).

———. *Virtue, Commerce and History: Essays on Political Thought and History, Chiefly in the Eighteenth Century*: (Cambridge University Press, 1985).

———. *The Rise of Free Trade Imperialism: Classical Political Economy, the Empire of Free Trade and Imperialism 1750-1850*, (Cambridge University Press, 1970).

Silver, A: 'Friendship in Commercial Society: Eighteenth-Century Social Theory and Modern Sociology', *The American Journal of Sociology*, Vol. 95, No. 6 (May, 1990), pp. 1474-1504.

Skinner, Q: 'Meaning and Understanding in the History of Ideas', *History and Theory* 8 (1969), pp. 3- 53.

Slack, P: 'Material progress and the challenge of
affluence in seventeenth-century England', *Economic History Review*, 62, 9 (2009), pp. 576-603.

*The Cambridge History of Eighteenth-Century
Political Thought*, M. Goldie and R. Wokler
(eds.), (Cambridge University Press, 2006).

The Economic Limits to Modern Politics, John Dunn
(ed.), (Cambridge University Press, 1990).

The Varieties of British Political Thought 1500-1800,
J. G. A. Pocock, G. J. Schochet and L. G. Schwoerer (eds.), (Cambridge University Press, 1993).

Wealth and Virtue: The Shaping of Political Economy in the Scottish Enlightenment, I. Hont and M Ignatieff (eds.), (Cambridge University Press, 1983).

Winch, D: *Malthus*: (Oxford University Press, 1987).

———. *Riches and Poverty: An Intellectual History of Political Economy in England 1750-1834*, (Cambridge University Press, 1996).

Further information on the themes discussed in this book:

Is the Human Species Special? Why human-induced global warming could be in the interests of life (2010)

An Evolutionary Perspective on the Relationship Between Humans and Their Surroundings: Geoengineering, the Purpose of Life & the Nature of the Universe (2012)

Saviours or Destroyers: The relationship between the human species and the rest of life on Earth (2012)

http://neilpaulcummins.blogspot.com